I0829110

THE

NICOLSON PAVEMENT,

Invented by

SAMUEL NICOLSON,

OF

BOSTON, MASS.

[PATENT RIGHT SECURED.]

SECOND EDITION WITH ADDITIONS.

BOSTON:
PRINTED BY HENRY W. DUTTON AND SON,
33 and 35 Congress Street.
1859.

Putting down the Nicolson Pavement, at the crossing of Washington and Clark Streets, Chicago, Illinois.

THE NICOLSON PAVEMENT.

STATEMENT.

For nearly thirty years I was Treasurer and Superintendent of the Boston and Roxbury Mill Corporation, and had in charge the preservation of its nine miles of avenues and roads. The main avenue is generally known as the Mill Dam; but that unpretending name designates, in point of fact, a continuation of Beacon street, one of the most beautiful avenues of the city, and which connects Boston with the city of Roxbury and its immediate vicinity.

The frequent repairs of the avenue, made necessary by the travel to which it is exposed, led me to consider the general subject of paving. It seemed to me that the prime objects to be attained in paving, were, safety to travellers, horses and vehicles; durability; comparative noiselessness; healthfulness as to materials used; absence of dust, and exemption from rapidly accumulating and slippery mud. These essential elements, I fancy, are combined in what is now known as "The Nicolson Pavement."

Most of the pavement in Boston consists of stones collected on the beaches of the main land and of the islands. They have become somewhat scarce; but if abundant, they could not now be considered as the most unexceptionable material. Square blocks of stone have been used to a considerable extent; but as they wear smooth from attrition, they afford an insecure footing for horses, and are liable to grave objections. Iron has been used to a limited extent; but it is very expensive. Wood, in some instances, has been resorted to; but, heretofore, it has not answered the purpose. Yet, on reflection, I did not hesitate to adopt *wood* as the basis of my plan, preserving and combining it with other materials, in the modes hereinafter described.

THE METHOD.

which I now proceed to describe, is that which I first put into use. Early in the month of July, 1848, I prepared a section of about one hundred feet in length, and of the width of the road, at the toll-house, on the Western Avenue of the Boston and Roxbury Mill Corporation, near Beacon street. I first caused the road to be properly graded, and then covered by a composition of coal tar, sand and lime, about two inches in thickness, to prevent the injurious effects of moisture under the surface. On this composition or concrete foundation, I laid down four different modifications of my wooden pavement.

FIRST MODIFICATION.

The following diagram illustrates this first modification: —

No. 1.

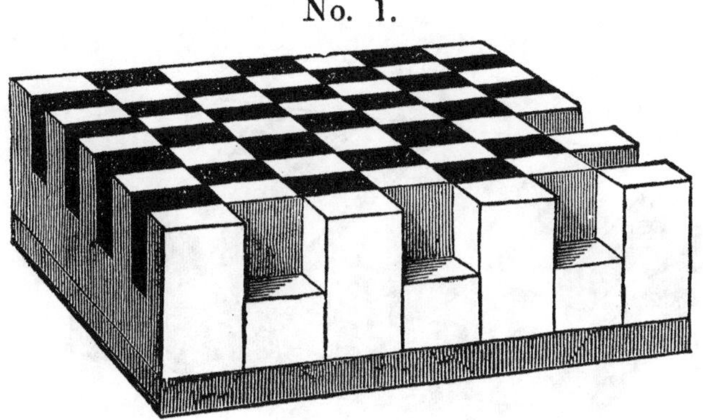

On about fifteen feet square of the road so prepared, I placed spruce blocks three inches square, in alternate lengths of eight and four inches, in a vertical position, so as to present a checkered surface. The spaces above the short blocks were then filled up to the level of the long blocks, with coarse pebbles, or small pieces of broken stone. After being rammed, boiling pine tar was poured over the whole surface, in order that it might penetrate all parts of the pavement, and into the fibrous structure of the wood itself. About one inch of dry sand was spread over the surface, and rammed. The blocks of this square were all tree-nailed (pinned) together, to prevent them from becoming uneven by the weight of heavy teams.

SECOND METHOD.

The next diagram illustrates the second modification: —

No. 2.

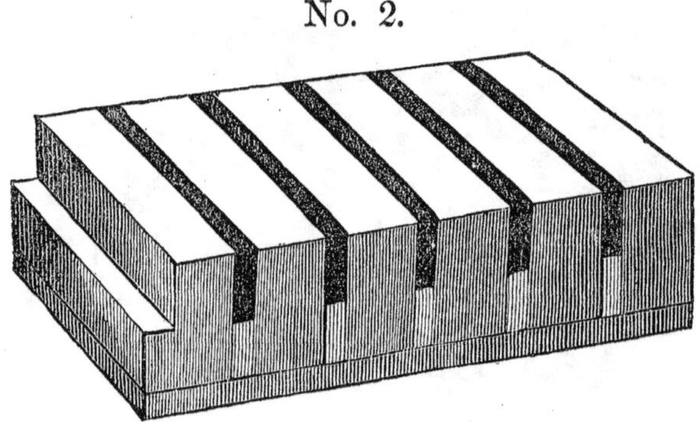

In the second modification, the short blocks were omitted, and the long blocks placed endwise, and next to each other, so as to present latitudinal ridges or ranges, about one inch apart; the ridges were kept separate by the introduction of strips of boards one inch thick, by about four inches in width; on the top of which, between the ranges of blocks, broken stone or pebbles were rammed, and then the hot tar and sand applied as in the first modification.

THIRD METHOD.

The third modification consisted in the use of round spruce blocks, eight inches in length. These were placed upon the composition first mentioned, so as to present small openings between them, which were filled as in the other instances. These blocks were used as cut from the tree; that is, the bark was retained, and the sap was not extracted.

This diagram illustrates the third modification.

No. 3.

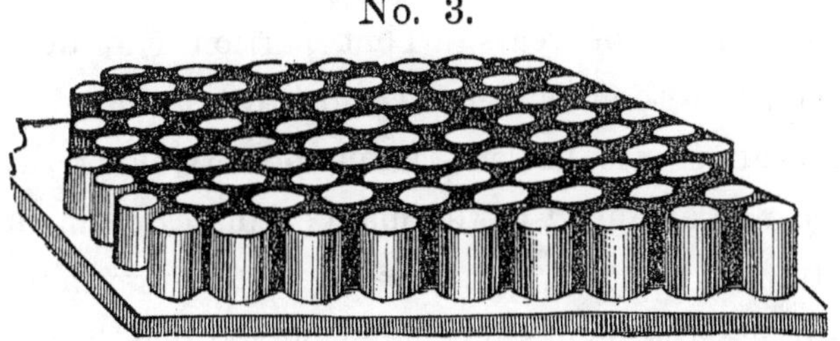

FOURTH METHOD.

The fourth modification is like the first, but with this difference; the blocks were not tree-nailed together, but their firmness left to depend upon the solidity of the base, and the composition introduced between the blocks.

The road, so constructed, was opened for travel on the 19th of July, 1848; and *for seven years it neither received nor required the slightest repair.*

After the pavement had been down about five years it became necessary to remove some of the blocks, in order to lay the pipes of the aqueduct. These blocks were found to be in excellent preservation. Some of them were exhibited at the rooms of the mayor and aldermen, and some at my office, where they might be readily seen and examined.

Notwithstanding the long exposure of the pavement on the Western Avenue, to the changing influences of frost, moisture and dryness, it remained in good order for years. That portion laid down according to the first, second and third of the above modifications was in remarkable preservation; that which was arranged in the manner indicated in the fourth modification, presented a slight unevenness of surface, the blocks not being pinned together. I have, therefore, adopted the plan of placing a flooring of boards upon the first composition used after grading the road, and it promises to keep the blocks at an even surface for many years.

In the spring of 1853, the mayor and aldermen of this city ordered the southerly half (about two hundred feet) of the bridge, leading from Sea street to South Boston, to be prepared with my pavement. I adopted the second modification, that of the latitudinal ridges, and it remained until the bridge was broken up, although it was exposed to the incessant and heavy travel of this thronged and crowded avenue from the city proper.

In the autumn of 1853, it was ordered that Mason street should be paved with my pavement; and in May, 1854, the board of mayor and aldermen renewed the order. This street is between Washington and Tremont streets, running parallel with them, and in the rear of the Boston Theatre. It was paved in ridges, similar to the method used on the bridge to South Boston; but the foundation was made by laying on the street, when graded, a preparation of coal tar and sand; on this was placed a covering or floor of one inch boards, upon which the blocks were arranged. The filling, between the ridges, was of sand, and the whole was then covered with hot coal tar, hardened with resin. Sand does not receive the tar and solidify as satisfactorily as pine tar or thin asphaltum combines with pebbles or broken stone, and I do not recommend it if the other materials are at hand.

The travel in Mason street was very inconsiderable and light before my pavement was laid; since then it has increased twenty fold. Teams and heavy wagons which used to pass over the stone pavement of Tremont street, while going to or returning from the Providence Railroad depot, afterwards sought the safe and easy pavement of Mason street.

On the application of the owners and occupants of real estate situated on West street, (from which runs Mason street,) and on Exchange street, the board of mayor and aldermen consented to the removal of the stone pavements from those streets, and the substitution of my pavement. Exchange street runs from Dock square to State street, and, like the bridge before mentioned, is one of the most crowded thoroughfares of the city. Perhaps no street could be found in Boston in which the travel is so trying to the pavement. The pavement was not of a uniform character through the extent of the street; I preferred a diversity of manner as to wood, to test the different modifications, and also the qualities of different woods. The first one hundred and thirty feet of the street, starting from State street, were paved with round spruce blocks, as in Diagram No. 3, (not divested of bark or sap,) eight inches long, and from three to five inches in diameter. For the next eighty feet, plank blocks, as in Diagram No 2, were used; they were three inches thick; between the latitudinal ridges were placed strips of board, one inch thick and three inches wide. On the next portion of the street were used round spruce blocks, four inches in diameter, with the bark on. On the remaining portion, plank blocks were used, as in the second portion.

The street was first graded, the surface rammed, and then a covering of melted asphaltum spread over the whole street, about the half of an inch in thickness, which was covered with a thin coat of sand. The round blocks were loosely placed upon the sand, and the spaces between them filled with coarse pebbles and rammed; asphaltum, heated to a high temperature, was poured on, so as to enter into the wood and among the pebbles, rendering the whole firm and solid. In the same manner, the whole pavement was treated with the pebbles and hot asphaltum.

At the termination of the first one hundred and eighty feet, the asphaltum spread upon the ground was covered with a flooring of hemlock boards, as I was of opinion, that the surface of the pavement would thus be better preserved against the effects of the incessant and very heavy travel of Exchange street. After the pavement was finished, a thin covering of hot asphaltum, mixed with sand, was applied, and the street opened for travel.

West street was the next to be paved. I proposed, after the usual grading of the street, to cover it with a composition to protect the pavement from moisture; then to lay a board or plank flooring, and to use spruce blocks six inches instead of eight inches long; to fill the interstices as before, but covering the whole with hot pine tar, or a composition softer than the asphaltum in Exchange street.

Such, then, are substantially the methods to be employed. Slight variations in the use of materials, or in their arrangement, may be adopted, as experience may suggest; but there is the principle, and I believe it to be philosophical and practical.

The method of paving in ridges transversely to the street, as delineated in Diagram No. 2, will be found to be the best for such great thoroughfares as Washington street in Boston, and Broadway in New York. In streets where stone is used for paving, it will be seen that the imperfect joints are soon worn into extensive longitudinal grooves, thus causing an uneven surface, and augmenting the danger to horses and vehicles.

On the Western Avenue, where the pavement laid for seven years, and on South Boston Bridge, where it had been in use for a year and a half, there was not the slightest indication of longitudinal grooves or gutters, or any inequality of surface.

This method is especially adapted to railroad and other bridges. For bridges, the blocks may be but four inches long, which will save half of the usual cost of materials. There will be no earth beneath the blocks, and as the risk of moisture will be thereby diminished, the durability of the pavement will be insured for many years, while the mechanical construction will add greatly to the firmness and solidity of the bridge. For the tracks between the rails of a railroad on which horse-power is used, the adaptation of this method appears to be perfect; for while it gives a secure foot-hold to the horse, it presents a clean and level surface for cross travel, and as it is less hard than stone, horses will not so soon lose their usefulness.

I proceed, briefly as possible, to enumerate and describe some of its important advantages.

SAFETY.

The security of travellers, horses and vehicles, is of transcendent importance. I think I have effected this in an eminent degree. I have never heard of an accident caused by the slipping of horses, on any of the pavements which I have laid. The small prominences of the wooden blocks so fit the foot of a horse, that, at every step, he catches upon, or is sustained by, the resisting properties of the composition rammed into the interstices. Even the fibres of the wood become tough, and present a resisting rather than a smooth surface, from the insinuating property of the coal tar or asphaltum covering, mixed up as it is with fine portions of earth and sand.

Horses are spared the great efforts they are often obliged to make in starting a heavy load, from the superior evenness of the surface, and yet they obtain a safe foothold, which they continually fail to do, on the round or square stone pavements.

NOISELESSNESS.

Total exemption from noise would be no recommendation of a pavement, for obvious reasons. A mitigation of it is all that can be desired. The offices in Exchange street were highly objectionable, if on the front, by reason of the excessive noise incident to the immense travel in this narrow street. To remedy the evil was one of the motives for the substitution of my pavement. The relief is very great. It needs no labored paragraph to show that the comparative noiselessness of a pavement, in a populous city, is an object of no inconsiderable importance. My pavement is as free from noise, as is consistent with the safety of the community.

CLEANLINESS.

Observation has abundantly proved, that dust does not accumulate so rapidly or remain so long as on the ordinary stone pavements, which is a great relief both to travellers and to those who reside on the street. The superior surface drainage of my pavement prevents the continued formation of mud, which is slippery and dangerous to man and beast, and always offensive. The pavement is too tight to admit the entrance of water, and as the composition resists the sub-moisture, the result is something over and beyond a mere clean thoroughfare; it becomes a preventive to foul exhalations, and a preservative of the public health.

Hence my invention may be introduced into south and southwestern cities with decidedly beneficial results, in a sanitary point of view. In cities exposed to *miasmatic vapors*, the pavement, by reason of its compactness, and the bituminous materials entering into its composition, cannot but have a purifying effect upon the surrounding air. There can be no exhalation from it, at any time, particularly after rains, and under the rays of a hot sun, which will *not* be salubrious, rather than pernicious.

DURABILITY.

It is not pretended that this pavement is indestructible; but in durability it will favorably compare with any now in use. The substitution of hard wood would tend to the permanency of the pavement; but it would involve a greater expense, and I have used the softer kinds for that reason. So, too, if the latter were subjected to the several well known chemical processes devised for the preservation of wood, greater permanence would be obtained. But I have not resorted to such means, because I have preferred to use common materials, in this particular, for prudential considerations, as well as to test the preserving qualities of the composition.

It is found, that the surface of the blocks of wood in the pavement on the Western Avenue and on South Boston Bridge, has become case-hardened, as it were. The blows from the feet of horses have slightly broomed or opened the fibres of the wood, into which sand has entered and been hardened down by pressure of wheels, and thus largely contributed to the preservation of the pavement, and avoided the necessity of any repairs. Seven years are proof of its comparative durability, even when constructed of soft wood. Frost does not heave it, nor has the heat of summer impaired it.

ECONOMY.

A pavement may be somewhat costly, and yet, in a long series of years, be found to be an economical one, because there enters into the construction of the best possible pavements, elements aside from the first cost. On this head, however, I intend to confine myself to figures.

I can furnish my pavements, in this city, on due notice, so as to have time for collecting the materials, at a cost of two dollars, or a trifle less, per square yard. That on South Boston Bridge cost a little over two dollars. This is much less than the stone blocks cost, and seventy-five cents per square yard more than common round or cobble stone pavement. The stone cube pavement, in a greatly travelled street, will only last about five years without repairs. It is very noisy and slippery. The pebble pavement is also noisy, dirty, becomes full of holes, and requires repairs, in streets like Exchange street, in less than two years. Judging by the past, my pavement must be considered as more economical than any now in use in this city.

Such then are the qualities of the proposed method. It has successfully stood the severe trials to which it has been subjected. Not until its merits had been tested have I ventured to recommend its adoption. I have patiently waited until I could present it with the stamp of approbation which time has impressed upon it. The favorable result of all experiments, and the approbation of scientific and official gentlemen, justify me in the belief, that I offer to the public a safe, durable, and economical pavement.

SAMUEL NICOLSON.

Boston, *November*, 1859.

POSTSCRIPT.

Corporations, individuals and other parties desiring further information concerning the pavement, or wishing to purchase an interest for any City, State, or District, may address me,

AT OFFICE,

No. 8, *Phœnix Building*,

BOSTON.

☞For new Testimonials, and a design for an economical Street Railroad in combination with the Wood Pavement,—refer to pages 25 and 32.

TESTIMONIALS.

I have concluded to append a few testimonials from distinguished sources, in favor of my invention. Emanating, as some of them do, from gentlemen who have been or are connected with the City Government, or who are distinguished for scientific attainments, they are entitled to the highest consideration.

Letter from Dr. C. T. Jackson, State Assayer, &c.

Boston, Dec. 5th, 1854.

Samuel Nicolson, Esq.

Dear Sir,—Having examined specimens of your patent wood and pitch pavement, on the causeway of the Boston Mill-Dam, and on South Boston Bridge, as also the more recent improvements you have made in this kind of pavement in Mason and Exchange streets, and having made due inquiry of persons residing on the streets and bridges above mentioned, as to the actual wear of the new roads paved by you, and the influence of heat, cold, and water upon the said roads so paved, I am prepared to state, that your method has been proved to be a valuable improvement, of great importance to the community. It is obvious that your pavement is, while firm enough to support the heaviest carriages and loaded wagons, the most noiseless and cleanly, giving but little if any dust, and forming no slippery mud.

It is so tight as to prevent the entrance of water, and therefore is not liable to be heaved by frost, against which your mastic, or cement, also acts, being an imperfect conductor of heat.

In reply to your question, as to the effect of this kind of pavement in southern and western cities of the United States, in preventing the emanation of miasmatic vapors from the soil, I would state my confident belief, that by shutting off the water from the surface of the street, and by preventing exhalations caused by the action of the sun's heat on wet soils, or earth and vegetable matters, your pavement will act as a highly sanitary means, wherever it is extensively employed in paving the streets of southern and southwestern cities. It will evidently protect the cities so far as it can be made to cover up the ground, from which miasmatic vapors arise.

From careful inquiry, as to the effects of solar heat in softening the cement, I find that there is little danger to be apprehended from that source, since I learn that during the hottest weather, in this city, the cement did not soften at all. The large mixture of earthy matters prevents the pitch from running or becoming adhesive to the feet. I learned, also, that there was no trouble from the slipping of the feet of horses in travelling over your road; the breaks of the surface, contrived by you, serving to give the horses good foothold.

Having seen this pavement, that had borne the travel on the Mill Dam road for six and a half years, I am satisfied that the road is reasonably permanent, and will not require relaying more frequently than stone pavements usually do.

Respectfully, your obedient servant,

CHARLES T. JACKSON, M. D.,

Assayer to the State of Massachusetts and to the City of Boston.

Letter from Dr. Hayes, State Assayer.

To Samuel Nicolson, Esq.

Dear Sir,—I have been acquainted with your pavement for some years, and am prepared to state, that it includes some important adaptations, strongly recommending it to public use.

The mechanical arrangement of the blocks is such, as to secure freedom from noise and risk of slipping by horses; surface drainage is also perfectly effected; the wooden blocks being quickly dried after exposure. By the use of the bituminous composition, or mastic, the durability of the wood is preserved, the masses becoming hardened by impregnation, so that the exhalations from rot are not present. Indeed, the vapors which might exhale under the rays of the sun, are those which neutralize miasmata, if existing in the air. The gravel so perfectly unites with the mastic, that a kind of rock is produced, specially adapted for roads over porous or imperfectly drained soils.

Having a favorable opinion of this invention, I hope it may come into extended use.

Respectfully,

A. A. HAYES, M. D.,

Assayer to State of Massachusetts.

16 *Boylston Street, 4th Dec.*, 1854.

Letter from the Hon. J. V. C. Smith, Mayor of Boston.

Mayor's Office, City Hall, *Boston, Dec. 7th,* 1854.

Dear Sir,—Having examined your new mode of paving, I take pleasure in stating that it promises to meet all reasonable expectations in regard to durability, economy, safety, and cleanliness. The severe trials to which it has been exposed on the Boston and Roxbury Mill Dam, on the North Bridge leading to South Boston, and on Mason street, have tested its qualities and established its value. It is a great improvement and merits success.

Very respectfully, I have the honor to remain,

Your friend and servant,

J. V. C. SMITH, *Mayor.*

Samuel Nicolson, Esq., Boston.

Letter from Josiah Dunham, Jr., Esq., Alderman of the City of Boston, and one of the Committee on Paving.

Boston, December 7th, 1854.

My Dear Sir: Being a resident of South Boston, but in the daily habit of visiting the city proper in the discharge of my official duties as a member of the Board of Aldermen, I have had opportunity to watch the useful and enduring qualities of the pavement laid by you on the southerly portion of the bridge leading from Sea street. Whatever misgivings I may have had in the beginning as to its success, have been entirely removed by the result. It has afforded a safe and easy road for the incessant travel over this crowded avenue, without the slightest outlay for repairs. I entertain the highest opinion of the merits of your invention, and hope to see it very generally introduced.

Very respectfully, your obedient servant,

JOSIAH DUNHAM, Jr.,

On the Committee on Paving.

Samuel Nicolson, Esq., Boston.

Testimonial from Members of the Board of Aldermen.

The undersigned, members of the Board of Aldermen of the city of Boston, having had occasion in the course of their official duties to notice the construction and safety of the pavement invented by Samuel Nicolson, Esq., are satisfied, from the experiments which have been made, that it possesses the important characteristics of the best method of paving, and is an improvement of high value.

Its economy, safety, and reasonable durability, entitle it to the most fovorable consideration.

GEORGE F. WILLIAMS,
TISDALE DRAKE,
GEO. ODIORNE,
B. L. ALLEN,
W. WASHBURN,
JOHN T. DINGLEY.

Boston, December 7th, 1854.

Testimonial of the Hon. Benjamin Seaver, late Mayor of Boston.

While mayor of this city, my attention was directed to the "Nicolson Pavement," which had been laid, some years previous, on a portion of the main avenue of the Boston and Roxbury Mill Corporation. On examination, the Board of Aldermen and myself were so impressed with the conviction of its value, as a safe and durable pavement, that, in the spring of 1853, an order was passed that about two hundred feet of the southerly half of the bridge leading from Sea street to South Boston, should be so paved.

It has been exposed to the continuous stream of heavy travel which flows over that thoroughfare, and to the different temperatures of the successive seasons of the year, without the necessity of any repair whatever.

In my judgment, it is an agreeable, safe, and economical pavement.

BENJAMIN SEAVER.

Boston, November 27, 1854.

Testimonial of Thomas P. Rich, Esq., late member of the Board of Aldermen.

During the year 1853, I was, as one of the Aldermen of the city of Boston, Chairman of the Committee on the "Paving and Repairs of Streets."

My official duty led me to an examination of the different methods of paving. I had formed a favorable impression in regard to the "Nicolson Pavement," in consequence of the durable character of that which had been laid, some years before, on the Western Avenue. I concurred in the laying it on the southerly portion of the bridge leading from Sea street to South Boston, and the result answers my expectations. I also offered the order for paving Mason street. I entertain a high opinion of the pavement, by reason of its durability, exemption from noise and mud, of its safety to horses, and as a secure and economical pavement.

THOMAS P. RICH,
59 Beacon Street.

Boston, December 7, 1854.

The undersigned, residents of South Boston, or its vicinity, having occasion to pass over the bridge leading to Sea street, almost daily,—have noticed, with great satisfaction, the beauty, security and durability of the "Nicolson Pavement," which is laid down on the southerly portion of the bridge.

Considering the incessant travel over this thronged avenue—much of which is by the heaviest teams, carting iron to and from South Boston; and that a large proportion of the Quincy Granite, in loads of many tons each, passes over this bridge; considering also, that during the eighteen months this pavement has been laid down, no perceptible wear of its surface, or displacement, from frost or otherwise, appears; that it affords an admirable footing for horses; exhibits great evenness of surface; and promises durability, without requiring frequent and expensive repairs—and, understanding that its original cost is moderate, we are of opinion that its general introduction is highly important and desirable.

CYRUS ALGER,
EBENEZER A. LESTER,
HENRY WASHBURN,
ALGER & REED,
THOS. THACHER, JUN.,
B. T. REED,
Treas. Bay State Iron Co.
WM. B. DORR.

Boston, December 11, 1854.

It is hardly necessary to say, that the preceding certificate is signed by gentlemen of high standing in this community, representing manufacturing establishments at South Boston, whose operations are not exceeded by any engaged in similar business, in this part of the country. The following authorized estimates of the amount of teaming to and from these establishments, (and some others,) for the last year, give quite an imperfect idea of their number and the magnitude of their business, and but faintly exhibit the industry, enterprise and prosperity of South Boston. They are introduced for the sole purpose of illustrating the character of a portion of the travel to which the pavement laid down by me on the bridge leading from Sea street to that important section of the city, has been constantly subjected.

The railroad iron, carted in loads of from four to nine tons each, by the Bay State Iron Company, is estimated at about -	25,000 tons.
The iron carted to and from the Wire Manufactory, at	10,000 "
The iron and spikes carted for the Spike Manufactory,	3,000 "
The iron and coal carted for the Foundry of Messrs. Alger & Reed, at - - - - - - -	6,000 "
The iron and coal carted for the South Boston Iron Company, estimated to be over - - - - -	3,000 "
The pig iron carted for the Fulton Iron Company, at	4,500 "
The truckage from the manufactory of Rainstead, Dearborn & Co., at - - - - - - -	3,000 "

By the Weymouth Iron Company, twenty-five thousand casks of nails were teamed into the city, and an equal weight of iron was taken back to the manufactory, during the past year.

The amount of granite annually conveyed over said bridge, from the quarries at Quincy, is estimated at over one hundred thousand tons—a load of which averages from four to over eight tons.

The freight trucked to and from the city, over this bridge, by the Old Colony Railroad Corporation, ranges from thirty-five thousand to forty thousand tons per year.

Mr. Caleb Thurston estimates the amount teamed by him over the same bridge, during the past year, at over eight thousand tons; his loads are mostly of iron, coal, machinery, and locomotive engines; the last, it is well known, average over fourteen tons each.

To prevent misapprehension, it should be added, that most of the coal and iron used by the manufactories at South Boston, is conveyed there by water, and that the object of the above statements is, not to present even an approximate estimate of the aggregate of their business, or that of South Boston, but merely to show, by way of inference, the enduring qualities and value of the "Nicolson Pavement," for great thoroughfares and bridges.

S. N.

In addition to the testimonials contained in the first edition, I insert a few letters from official and other gentlemen at the West, to which I refer; and, particularly, to the communications from Chicago, Illinois, where the pavement is in high favor, and where more than *thirty thousand* square yards have been laid down, as will be seen on reference to the letter of R. Cleveland, Esq., Superintendent of Public Works in that city.

Letters from Prof. John C. Beck, of Cincinnati College.

87 *Broadway, Cincinnati, O., July* 20, 1859.

Mr. Samuel Nicolson.

Dear Sir,—I would like to read your pamphlet on pavements. I usually deliver two lectures on the health and comfort of communities each session, and always have something to say about pavements.

Sanitary regulations should induce our cities to abandon those old methods of bouldering and paving streets.

Very respectfully,

JOHN C. BECK, M. D.,

Professor of Medical Jurisprudence of the Cincinnati College of Medicine and Surgery.

---

87 *Broadway, Cincinnati, O., Sept.* 2, 1859.

Mr. S. Nicolson.

Dear Sir,—With pleasure I received your very kind favor of the 26th ult., and in reply would say that I have not the slightest doubt of the superiority of your pavement over all others to which I have directed my attention. Its adoption would certainly lessen, very materially, the mortality of our cities, and add essentially to the average longevity of residents.

Very respectfully,

JOHN C. BECK, M. D.,

Professor of Medical Jurisprudence of the Cincinnati College of Medicine and Surgery.

Copy of a Paper presented to the Mayor and Aldermen of the City of Boston.

To the Hon. Mayor and Aldermen of the City of Boston:—

The undersigned, owners of real estate, and residents on Washington street, between Court street and Winter street, having learned that it is your intention to remove the stone block pavement, now in said street, would respectfully and particularly call your attention to a substitution of the "Nicolson Pavement," which, in our opinion, has been most successfully tested for several years on some of the streets and avenues of this city, as possessing qualities superior to any now in use, viz., its noiselessness, safety to horses, cleanliness and economy, and particularly, that *it effectually resists frost and water, &c.*

Signed by, viz.:—

George W. Warren & Co.,
Adams & Co.,
John F. Pray,
Wm. D. Ticknor & Co.,
Hermann & Co.,
P. Fowle & Sons,
Stimpson & Co.,
Eayrs & Fairbanks,

James Munroe,
E. H. Wade,
Lincoln & Foss,
Shorey & Co.,
Sumners & Co.,
Call & Tuttle,
Abraham Hews, Jr.,
Augustus H. Hews,
And about sixty others.

Boston, February 23d, 1855.

The pavement above petitioned for was laid in 1855.

Copy of Recommendation from the above Petitioners.

Boston, August 29, 1859.

The undersigned, whose names are on the petition to the mayor and aldermen of the city of Boston, for laying the "Nicolson Pavement" on Washington street, hereby again express our preference for that pavement, and recommend its adoption where a quiet, safe, clean and economical pavement may be wanted.

Adams Express Co.,
(A. Adams, Sup't.)
Sumners & Co.,
E. H. Wade,
Call & Tuttle,
Chase & Shorey,
Hermann & Co.,
P. Fowle & Sons,
A. H. Hews,
Eayrs & Fairbanks,
Lincoln & Foss,
W. D. Ticknor & Co.,
James Munroe & Co.,
Geo. W. Warren & Co.,
And others.

Letter from A. P. Turner, Superintendent of Streets.

Paving Department,
Superintendent of Streets' Office,
City Hall, Boston, Aug. 10, 1859.

To S. Nicolson, Esq.

Dear Sir,—In answer to your request I will state that there is now down, in this city, the "Nicolson Pavement," on Mason street, part of West street, part of Chauncey street, and part of Exchange street. Of the durability of this pavement I refer to my annual report to the city government in January last. Yours, &c.,

ALFRED P. TURNER,
Street Superintendent.

COPY OF REPORT.

"The Nicolson Pavement has many good qualities which it possesses over others, such as its freedom from earthy exhalations, and its peculiar properties, preventing, in a degree, the formation of frost beneath it, &c.

"It might I should think be adapted to southern and southwestern cities on account of the peculiar properties mentioned, rendering it sanitary in its character, and because wood is plenty there, while stone at all suitable is very difficult to be procured. Whenever wood pavement is placed down, I believe the hard wood blocks, similar to those on Exchange street, would better meet the wishes of the advocates of this method of paving, &c.
A. P. T.

Letter from Dr. Hayes, State Assayer.

16 *Boylston Street, Boston,* 10*th Aug.*, 1859.

To Sam'l Nicolson.

Dear Sir,—You ask me to state what surfaces are now paved with your patented improved pavement. The streets under my constant observation are,
Mason street,
West street,
Chauncey street,
Exchange street,
partly in soft and partly in hard wood.

These streets invite, by the smoothness of their surface, the largest and heaviest traffic of our city. The heaviest loads drawn in any city are constantly, by night as well as by day, passing through these streets, and the duration of your pavement under such exposure surprises every intelligent person. It has borne more service than the iron and stone pavement (Terry's) could endure, and is a most desirable kind of protection.

The experience gained after the lapse of eleven years is decidedly in favor of hard wood for the blocks; and, it appears to me, that this kind of pavement is specially adapted to cities of warm climates on similar grounds, while comfort, cleanliness, economy and protection of the public health, are ensured by its use.

With high respect,
A. A. HAYES, M. D.,
Consulting Chemist and State Assayer.

Letter from William Parker, Civil Engineer.

Boston, August 10, 1859.

To S. Nicolson, Esq.

Dear Sir,—In compliance with your request I cheerfully give you the following statement respecting the Nicolson Pavement laid down in this city:—

Excepting the portion in Exchange street, and a small piece of which you informed me in Chauncey street, all the specimens here have been laid of soft and friable wood—chiefly spruce—and most of it has been subjected to a heavy wear—beyond what is usual in most other cities—notwithstanding which, it has endured for three years and upwards—showing now a condition decidedly preferable in my opinion to most boulder pavements—and more easily repaired.

The piece laid in hard wood, in Exchange street, has done nobly, and is now, after two winters, quite unaffected, and the best pavement in this city; confirming my often expressed conviction, that for all interior cities, your pavement, well laid in hard wood, is the best, and ultimately for all interests the cheapest pavement known. I am

Respectfully yours,

WM. PARKER, *Civil Engineer.*

Ex-President of Boston and Lowell Railroad.

Letter from Dr. C. T. Jackson, State Assayer.

State Assayer's Office,
Boston, August 11, 1859.

Samuel Nicolson, Esq.

Dear Sir,—I have examined your patent wooden pavement in numerous places in Boston, and can certify to its efficiency. A pavement of yours now exists in Mason street, over which I walk every few days, and which I always notice. It is in perfect condition and is a very nice pavement.

In the muddy soils of the Western States I think this pavement will prove of great value, for stones cannot be readily had there, while wood is abundant and cheap.

I have never heard that fault was found with this pavement when properly laid, and I know it has proved efficient and durable.

Respectfully, your obedient servant,

C. T. JACKSON, M. D.

State Assayer.

Letter from E. S. Chesbrough, Engineer, Chicago.

Chicago, Ill., August 11, 1859.

Dear Sir,—Yours of the 4th inst., after being opened by another, was handed to me. No one in this city enjoys the title, or unites in his office, all the functions that properly belong to a "city engineer."

About three years ago, I was, with others, consulted in relation to the propriety of laying down a piece of Nicolson pavement here. Having seen a good deal of it in Boston, I recommended a trial of it in Chicago. Others thought the same about it, and some two hundred feet was laid that year, and as much more the following season. No pavement in the city has given so much satisfaction as this; as it is very pleasant to drive over, being without disagreeable noise or jarring, continues in good order, has required no repairs, has been less expensive than other kinds to keep clean, and bids fair to last some years longer. The blocks are of soft pine, and have worn down a quarter of an inch, I understand, in two and three-quarter years. The traffic over it has been quite heavy.

This year they are laying about three-quarters of a mile of it on South Clark street, one of the most important thoroughfares in the city. The blocks used now are white-oak, adopted in consequence of the rapid wear of softer blocks on some of the narrow but much frequented streets of Boston. It is a great advantage to the Nicolson pavement (as well as to other kinds) to have the streets so wide as to prevent the teams from going in the same track, which is unavoidable when they are so narrow as to afford just room enough for two to pass.

The owners of property here pay for paving the streets; but the work is done under the supervision of a city officer.

The advantages of the Nicolson pavement have already been mentioned. One *feared* disadvantage is, that it may decay or wear out so soon as to make it more expensive in the end than other kinds; but this fear is diminishing. The only other disadvantage I can think of is, it is more trouble to repair perfectly than other kinds, whenever it is necessary to cut away the planking under it, to dig trenches for water or gas pipes, or sewers. Unless this is done well, it makes a bad place in the street; but there is no serious difficulty in doing it well.

The contract paid this year is —— per square yard, including six inches of gravel under the planking. Very low.

Any other items I may have omitted, will be furnished with pleasure, if in my power, as soon as you let me know what has been omitted.

Very respectfully, yours,

E. S. CHESBROUGH,

Chief Engineer Sewage Commissioners, Chicago, Ill.

JOHN GAGER, ESQ., Memphis.

Letter from S. S. Greeley, Civil Engineer.

Chicago, May 31*st*, 1859.

JOHN GAGER, ESQ.

Dear Sir,— * * The first piece of the Nicolson pavement laid here was put down in November, 1856, of white pine blocks six inches long. The street is, and always has been, in perfectly good repair since, and not one cent has been expended upon it for repairs—some of the blocks were taken up and examined last fall; they showed a uniform wear of three eighths of an inch in two years—this of white pine, which would naturally batter down nearly as much as that. In all of the pavements being laid here this year under this patent, hard wood (oak) is being used, which will conduce greatly to its durability, &c., &c.

S. S. GREELEY, *Civil Engineer.*

Letter from A. W. Gilbert, City Civil Engineer.

Cincinnati, June 18, 1859.

To Sam'l Nicolson, Esq.

Dear Sir,—I have just returned from a visit to Chicago, in company with the members of our city council and city authorities, and while there we examined and were much pleased with the pavement being laid down by Mr. Greeley upon your plan; and I am requested by several members of our council to write to you for the purpose of ascertaining upon what terms we can arrange with you to make the experiment upon a street in this city. If, after trying it, we find it an improvement upon our present mode, it will, I have no doubt, be quite extensively used here. I would like to hear from you as soon as possible.

Yours, &c.

A. W. GILBERT,
City Civil Engineer.

Certificate from Property Owners, Wells St., Chicago.

Chicago, August 17, 1859.

To whom it may Concern:

The undersigned, owners of property on Wells street, between South Water and Lake streets, hereby certify that the above-named block of Wells street was paved with the "Nicolson Pavement," with pine blocks, by Samuel S. Greeley; the north half in November, 1856, the south half in July, 1857.

No repairs have ever been done or required upon the pavement, and it is, so far as can be seen, in perfectly good order. Portions which have been taken up for the purpose of reaching pipes underneath, show no decay, and a uniform shortening of the blocks by about half an inch.

The pavement is dry, cleanly, and free from the noise of travel usual on stone pavements. This part of Wells street is a main thoroughfare, adjacent to a bridge, and has been constantly subjected to heavy travel.

Signed,

GEO. SMITH & CO.,
OGDEN, FLEETWOOD & CO.,
ALLEN ROBBINS.

Letter from A. W. Gilbert, City Civil Engineer.

Office City Civil Engineer,
Cincinnati, Sept. 8, 1859.

S. Nicolson, Esq.

Dear Sir,—Yours of 31st August is received, and I shall take pleasure in presenting your suggestions in regard to putting down horse railways on your pavement, to those interested here. I am in hopes Mr. Greeley of Chicago will feel sufficient interest in the matter to visit our city, with a view of laying down a portion of a street of your form of pavement. We should have some one well acquainted with the work to undertake it. I would not urge his coming, but would be glad to see him obtain a contract.

Yours, &c.,
A. W. GILBERT.

Certificate from R. Cleveland, Sup't Public Works.

Office of Sup't Public Works,
Chicago, Sept. 10, 1859.

I hereby certify, that a portion of Wells street, one hundred and fifty feet long, near the bridge, was paved in November, 1856, with the Nicolson pavement with pine blocks. After having been subjected to a very heavy travel for nearly three years, without repairs, it is still in perfectly good order, and looks likely to last several years longer. The pine blocks have been uniformly worn down about half an inch in the three years.

Most of the Nicolson pavement laid, since the first piece, has been constructed with blocks of white oak instead of pine. The total quantity now laid, and under contract for completion this fall, is 31,000 square yards, covering 1⅛ miles of streets and alleys.

The pavement is cleanly, noiseless, impervious to water and frost, affords an excellent foothold for horses, and is in all respects a very desirable pavement, both for occupants of adjacent buildings, and for persons driving over it. It is of great value in the saving of wear to vehicles, and of injury to horses.

Its durability has not been, as yet, fully tested here, but from the condition of the first piece laid down, after three years' use, it may be inferred to be nearly or quite as durable as the stone pavements in use here.

R. CLEVELAND,
Sup't Public Works.

A Design for the construction of an economical, quiet and durable Street Railway, in combination with the "Nicolson Pavement."

Horse railroads may be put upon a well-made "Nicolson Pavement," *without stringers and cross-ties.* The rails may be of common rolled iron, about two and a half inches wide and three-quarters of an inch thick, punched with holes to receive spikes or screws, of sufficient length to reach through the wooden blocks and into the floor boards of the pavement. The paving blocks under the rails should be reduced in length, just enough to bring the rails nearly flush with the street. In this manner, the action of the car wheels cannot spread, or change the guage of the track; thus rendering the construction of the road of a character to economize in repairs of both road and cars.

Beside the economy above named, a very important saving will be found in the lengthened term of usefulness in horses; as on common horse roads, where horses travel upon stone paving, they soon become lame and unfit for the business, while a paving of wooden blocks will be found easy to their feet, and will also lessen the repairs of horse-shoes.

SAMUEL NICOLSON,
No. 8 Phœnix Building, Boston, Mass.

Letter from Wm. Parker, Esq., Civil Engineer.

Boston, August 27, 1859.

Samuel Nicolson, Esq.

Dear Sir:—In reply to your suggestion this morning, that parties interested in Horse Railroad Tracks may also desire to use your pavement in combination therewith, I do not hesitate to express the opinion that the two systems are particularly well adapted to each other.

On a well-constructed "Nicolson Pavement" I have no doubt the iron may be directly laid without any wooden string pieces, and that for duration and the saving of repairs it would exceed all other modes hitherto known to me, a consideration which should go far in recommending the adoption of your pavement, though at enhanced first cost.

Respectfully, yours,

WM. PARKER, *Civil Engineer.*

Note.—Mr. William Parker is an Engineer of great experience, was formerly Superintendent of the Boston and Worcester Railroad, afterwards of the Baltimore and Ohio Railroad, and recently President of the Lowell and Boston Railroad.

www.ingramcontent.com/pod-product-compliance
Lightning Source LLC
LaVergne TN
LVHW011134110826
845150LV00008B/2332